ideal
school supply

MW01714940

Funtastic FROGS™

Addition & Subtraction

By Jill Osofsky

Contents

Published by ©Ideal School Supply
an imprint of
Frank Schaffer Publications®

Frank Schaffer Publications®

Ideal School Supply is an imprint of Frank Schaffer Publications.

Send all inquiries to:
Frank Schaffer Publications
8720 Orion Place
Columbus, Ohio 43240-2111

Funtastic Frogs™: Addition & Subtraction, Grades K–2

ISBN 1-56451-319-X

3 4 5 6 7 8 QPD 12 11 10 09 08

Teacher's Notes

This book is one in a series of books designed to develop children's mathematical thinking. Each book supports the use of Funtastic Frog™ Counters in activities that teach key mathematical concepts.

Funtastic Frog™ Counters are available from Ideal School Supply in six different colors and in three different sizes (3g, 6g, and 12g). A unique lacing feature allows children to use the frogs in a wide range of counting and patterning activities.

In *Funtastic Frogs™: Addition and Subtraction, Grades K-2,* children use the frog counters to:

- count on to add
- combine groups to add
- write addition sentences
- take away to subtract
- write subtraction sentences
- understand the meaning of addition and subtraction
- learn the vocabulary for addition and subtraction

The activities in this book are designed to teach addition and subtraction concepts using concrete objects. Children use the frog counters to act out situations that require combining groups to add or taking away parts of a larger group to subtract. They begin to connect the action of adding or subtracting to the words and symbols used to for the operations. They write simple addition and subtraction sentences. All the activities support current mathematics standards.

Contents

This book contains 24 activities divided into sections. Each section contains similar activities, giving the children the opportunity to practice and use each skill.

Sample solutions are provided on page 7. A photocopy master for the frog counters is included on page 8.

Math Skills and Understandings	Related Activities
Understand the meaning of addition and subtraction	Activities 1-24
Count on to add	Activities 1-5
Combine groups to add	Activities 6-15
Write addition sentences	Activities 11-15
Take away to subtract	Activities 16-19
Write subtraction sentences	Activities 20-24
Understand the vocabulary used for addition and subtraction	Activities 1-24

Suggestions for Classroom Use

The activities are sequenced by level of difficulty within each section and from section to section. Modify a section if you find it is too challenging for your children, or not challenging enough.

The activities can be introduced to the whole class using an overhead projector, the chalkboard, or sitting in a circle on the floor. Once children understand the directions for the activities, they can work in pairs, small groups, or individually at a learning center.

Encourage children to share their thinking with the whole class. Talking about their thinking and discoveries helps them clarify their thoughts and allows others to hear how they might solve the same problem a different way. Talking about how they solve problems helps children make mathematical connections and deepens their understanding.

Materials

You may wish to make an overhead transparency of the first activity page in each section as an introduction to the activities that follow.

For each child, pair of children, or group, you will need:

- pencils
- crayons or colored markers to match the colors of the frogs
- a tub of frogs in three sizes and six colors
- an activity sheet for each child or pair of children

If children are not able to record by writing numbers and coloring, make copies of page 8. The children can color and cut out the frogs, then paste or glue them in place to record.

Introducing the Activities

Before beginning instruction, it is important to allow time for children to freely explore the frog counters. This will allow them to satisfy their curiosity about the frogs before they use them in the activities. Encourage the children to compare the frogs and tell how they are alike and how they are different. Ask them to describe the sizes and the different weights. Ask questions to probe their thinking such as: *Can you show me five frogs? Can you show me five another way? Can you tell me how many large frogs you have? Do you have more small frogs or medium frogs?*

This exploration prior to beginning the activities provides an opportunity to introduce vocabulary used throughout the book. The frog sizes are referred to on the activity pages as *small, medium,* and *large.* Be sure all children become familiar with those terms before beginning instruction.

When you introduce the activities in a section, lead the children through the first activity in that section. Discuss the directions and how to record their work. When reviewing the activity as a group, encourage the children to talk about their discoveries. Make connections between the actions used for addition or subtraction and similar situations in the real world. Have children talk about their own experiences combining groups of objects, such as adding a new book or toy to a collection or their experiences taking away items from a larger number, such as taking one or two cookies away from a plate of six.

Mathematical Content by Section

Activities 1-5: Count on by one, two, or three. Children match and count the frogs on the page. They add one, two, or three frogs and count on to find how many in all. As you discuss the activities, emphasize the language used for the actions. To reinforce the concepts, encourage pairs to make up stories or other scenes using similar numbers and actions. They can exchange problems with a different pair and solve that problem by counting on.

Activities 6-10: Combine groups for sums to 10. Children match frogs to make two groups then combine the groups to find how many in all. Emphasize the language used for the actions. As reinforcement, students working in pairs can make up situations and stories to act out using similar number groups.

Activities 11-15: Use numbers and symbols to write simple addition sentences. Children learn to use the symbols for addition to write simple addition sentences.

Activities 16-19: Take away to subtract. Children take away or separate frog counters from the whole group to act out subtraction of numbers through ten. As they perform the actions, supply the vocabulary used to describe the actions for subtracting. Children can make up similar situations or scenes to act out subtracting for the basic facts through ten.

Activities 20-24: Use numbers and symbols to write simple subtraction sentences. Children learn to use the symbols for subtraction and write simple subtraction sentences.

Sample Solutions

These are sample solutions. Additional solutions are possible for some activities.

Activity

1 10
2 12

3A 5, 6
3B 7, 8

4A 4, 6
4B 6, 8

5A 6, 9
5B 8, 11

6A 3
6B 2
6C 5

7A 2
7B 4
7C 6

8A 4
8B 3
8C 7

9A 3
9B 5
9C 8

10A 5
10B 4
10C 9

11A 2+2=4
11B 3+1=4
11C 1+3=4

12A 3+2=5
12B 4+1=5
12C 2+3=5

13A 2+4=6
13B 3+3=6
13C 4+2=6

Activity

14A 4+3=7
14B 5+2=7
14C 3+4=7

15A 5+3=8
15B 2+6=8
15C 3+5=8

16A 2
16B 2
16C Answers will vary.

17A 2
17B 3
17C Answers will vary.

18A 4
18B 7
18C Answers will vary.

19A 4
19B 2
19C Answers will vary.

20A 7−1=6
20B 5−2=3
20C Answers will vary.

21A 6−2=4
21B 10−2=8
21C Answers will vary.

22A 9−6=3
22B 8−4=4
22C Answers will vary.

23A 10−5=5
23B 9−4=5
23C Answers will vary.

24A 9−8=1
24B 10−10=0
24C Answers will vary.

Name ____________________

Use small, medium,
and large frogs
Record your work.

Match each frog.
Count.

How many frogs in all? __________

Name ____________________

Use small, medium, and large frogs. Record your work.

Match each frog.
Count.

How many frogs in all? __________

Count on to add

ACTIVITY 3

Use small frogs.
Record your work.

Name ____________________

Match each frog.

A.

Count.
How many frogs?________

Add one more frog.
Count on.
How many in all?________

B.

Count.
How many frogs?________

Add one more frog.
Count on.
How many in all?________

Count on to add

Name ________________

Use small and medium frogs.
Record your work.

Match each frog.

A.

Count.
How many frogs? ________

Add two more frogs.
Count on.
How many in all? ________

B.

Count.
How many frogs? ________

Add two more frogs.
Count on.
How many in all? ________

Count on to add

ACTIVITY 5

Name ______________________

Use small frogs.
Record your work.

Match each frog.

A.

Count.
How many frogs? __________

Add three more frogs.
Count on.
How many in all? __________

B.

Count.
How many frogs? __________

Add three more frogs.
Count on.
How many in all? __________

Combine groups to add

ACTIVITY 6

Name ______________________

Use small and medium frogs
Record your work.

Match each frog.
Count.

A.

B.

How many frogs? __________

How many frogs? __________

C. Put all the frogs together.
Count.

How many frogs in all? __________

Combine groups to add

Name ____________________

Use small and medium frogs.
Record your work.

Match each frog.
Count.

A.

How many frogs? ________

B.

How many frogs? ________

C. Put all the frogs together.
Count.

How many frogs in all? ________

Combine groups to add

Name ________________

Use small frogs.
Record your work.

Match each frog.
Count.

A.

How many frogs? ________

B.

How many frogs? ________

C. Put all the frogs together.
Count.

How many frogs in all? ________

Combine groups to add

Use small frogs.
Record your work.

Name ______________________

Match each frog.
Count.

A.

B.

How many frogs? __________ How many frogs? __________

C. Put all the frogs together.
Count.

How many frogs in all? __________

Combine groups to add

ACTIVITY 10

Name ______________________

Use small frogs.
Record your work.

Match each frog.
Count.

A.

B.

How many frogs? ________ How many frogs? ________

C. Put all the frogs together.
Count.

How many frogs in all? ________

Write addition sentences

Name ______________________

Use small frogs.
Record your work.

A. Put 2 frogs on the  .

Put 2 frogs on the  .

Count how many frogs in all.
Write the addition sentence.

☐ + ☐ = ☐

B. Put 3 frogs on the  .

Put 1 frog on the  .

Count how many frogs in all.
Write the addition sentence.

☐ + ☐ = ☐

C. Use frogs to show 4 a different way.
Write the addition sentence.

☐ + ☐ = 4

Write addition sentences

Name ____________________

Use small frogs.
Record your work.

A. Put 3 frogs on the  .

Put 2 frogs on the  .

Count how many frogs in all.
Write the addition sentence.

□ + □ = □

B. Put 4 frogs on the  .

Put 1 frog on the  .

Count how many frogs in all.
Write the addition sentence.

□ + □ = □

C. Use frogs to show 5 a different way.
Write the addition sentence.

□ + □ = □

Write addition sentences

Name ______________________

Use small frogs.
Record your work.

A. Put 2 frogs on the  .

Put 4 frogs on the  .

Count how many frogs in all.
Write the addition sentence.

☐ + ☐ = ☐

B. Put 3 frogs on the 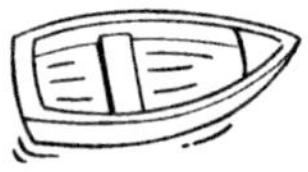 .

Put 3 frog on the  .

Count how many frogs in all.
Write the addition sentence.

☐ + ☐ = ☐

C. Use frogs to show 6 a different way.
Write the addition sentence.

☐ + ☐ = ☐

Write addition sentences

Name ____________________

Use small frogs.
Record your work.

A. Put 4 frogs on the 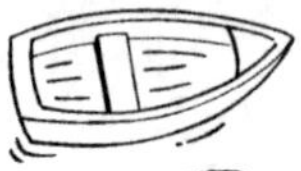 .

Put 3 frogs on the  .

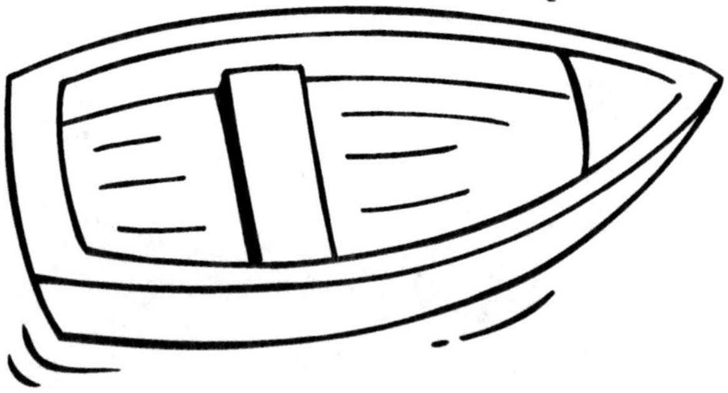

Count how many frogs in all.
Write the addition sentence.

☐ + ☐ = ☐

B. Put 5 frogs on the 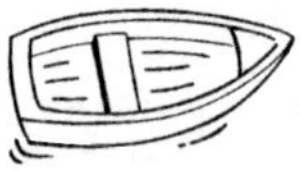 .

Put 2 frog on the  .

Count how many frogs in all.
Write the addition sentence.

☐ + ☐ = ☐

C. Use frogs to show 7 a different way.
Write the addition sentence.

☐ + ☐ = ☐

Write addition sentences

ACTIVITY 15

Name ______________________

Use small frogs.
Record your work.

A. Put 5 frogs on the  .

Put 3 frogs on the  .

Count how many frogs in all.
Write the addition sentence.

□ + □ = □

B. Put 2 frogs on the  .

Put 6 frogs on the  .

Count how many frogs in all.
Write the addition sentence.

□ + □ = □

C. Use frogs to show 8 a different way.
Write the addition sentence.

□ + □ = □

Take away to subtract

Name ____________________

Use small frogs.
Record your work.

Use frogs.
Act out each story.

A. There are 4 frogs on the  .
2 frogs hop away.
How many frogs are left on the  ? ________

B. There are 5 frogs on the  .
3 frogs hop away.
How many frogs are left on the  ? ________

Make up your own story.

C. There are ________ frogs on the  .

________ frogs hop away.

How many frogs are left on the ? ________

Take away to subtract

Name ______________________

Use small frogs.
Record your work.

Use frogs.
Act out each story.

A. There are 6 frogs on the .
4 frogs hop away.
How many frogs are left on the ? ________

B. There are 8 frogs on the .
5 frogs hop away.
How many frogs are left on the ? ________

Make up your own story.

C. There are ________ frogs on the .

________ frogs hop away.

How many frogs are left on the ? ________

Name ____________________

Use small frogs.
Record your work.

Use frogs.
Act out each story.

A. There are 7 frogs on the .
3 frogs hop away.
How many frogs are left on the ? ______

B. There are 9 frogs on the .
2 frogs hop away.
How many frogs are left on the ? ______

Make up your own story.

C. There are ________ frogs on the .

________ frogs hop away.

How many frogs are left on the ? ________

Take away to subtract

ACTIVITY 19

Use small frogs.
Record your work.

Name ____________________

Use frogs.
Act out each story.

A. There are 9 frogs on the  .
Take away 5 frogs.
How many frogs are left on the  ? ________

B. There are 8 frogs on the  .
Take away 6 frogs.
How many frogs are left on the  ? ________

Make up your own story.

C. There are ________ frogs on the 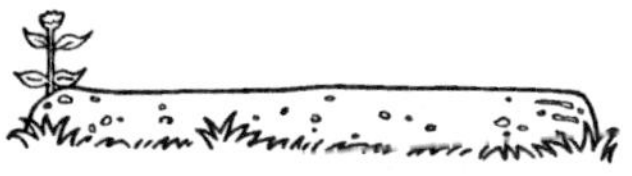 .

Take away ________ frogs.

How many frogs are left on the  ? ________

Write subtraction sentences

Name ____________________

Use small frogs.
Record your work.

Use frogs. Act out each story.

A. There are 7 frogs on the  .
1 frog hops off.
How many frogs are left on the  ? ______
Write the subtraction sentence.

☐ – ☐ = ☐

B. There are 5 frogs on the  .
2 frogs hop off.
How many frogs are left on the  ? ______
Write the subtraction sentence.

☐ – ☐ = ☐

Make up your own story.

C. There are ________ frogs on the  .

________ frogs hop off?

How many frogs are left on  ? ______
Write the subtraction sentence.

☐ – ☐ = ☐

Write subtraction sentences

Use small frogs
Record your work.

Name ____________________

Use frogs. Act out each story. Write the subtraction sentence.

A. There are 6 frogs in the 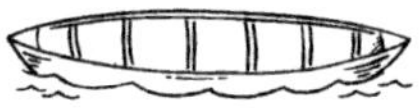 .
2 frogs hop out.
How many frogs are left in the 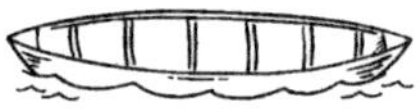 ?________

□ – □ = □

B. There are 10 frogs in the  .
2 frogs hop out.
How many frogs are left in the 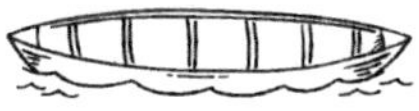 ?________

□ – □ = □

Make up your own story.

C. There are________frogs in the 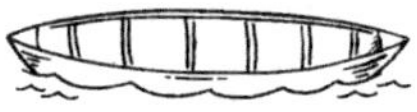 .

________frogs hop out.
How many frogs are left in the 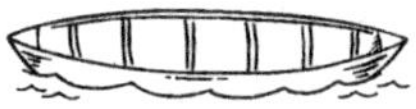 ?________

□ – □ = □

Use small frogs.
Record your work.

Name ____________________

Use frogs. Act out each story. Write the subtraction sentence.

A. There are 9 frogs in the  .
3 frogs hop out.
How many frogs are left in the  ? ________

□ – □ = □

B. There are 8 frogs in the  .
4 frogs hop out.
How many frogs are left in the  ? ________

□ – □ = □

Make up your own story.

C. There are ________ frogs in the  .

________ frogs hop out.

How many frogs are left in the  ? ________

□ – □ = □

Write subtraction sentences

Name ______________________

Use small frogs.
Record your work.

Use frogs. Act out each story. Write the subtraction sentence.

A. There are 10 frogs on the .
5 frogs hop off.
How many frogs are left on the ? ________

☐ – ☐ = ☐

B. There are 9 frogs on the .
4 frogs hop off.
How many frogs are left on the 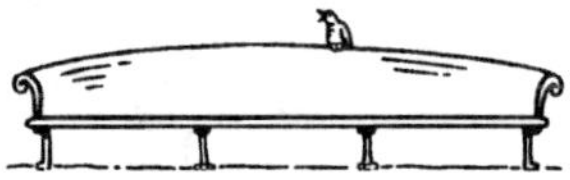? ________

☐ – ☐ = ☐

Make up your own story.

C. There are ________ frogs on the .

________ frogs hop off.

How many frogs are left on the ? ________

☐ – ☐ = ☐

Name ______________________

Use small frogs.
Record your work.

Use frogs. Act out each story. Write the subtraction sentence.

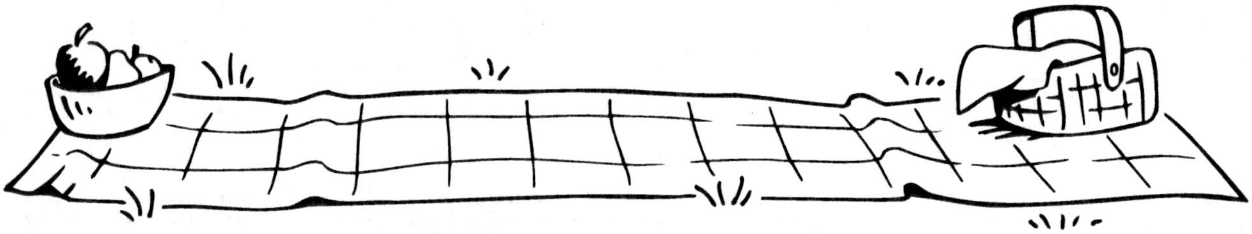

A. There are 9 frogs on the .
8 frogs hop off.
How many frogs are left on the  ? ________

□ – □ = □

B. There are 10 frogs on the .
10 frogs hop off.
How many frogs are left on the 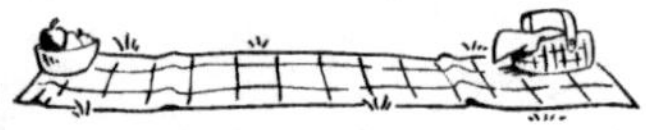 ? ________

□ – □ = □

Make up your own story.

C. There are ________ frogs on the .

________ frogs hop off.

How many frogs are left on the  ? ________

□ – □ = □

